Congressional
Research
Service

Prospects for Coal in Electric Power and Industry

Richard J. Campbell
Specialist in Energy Policy

Peter Folger
Specialist in Energy and Natural Resources Policy

Phillip Brown
Specialist in Energy Policy

February 4, 2013

Congressional Research Service

7-5700

www.crs.gov

R42950

CRS Report for Congress ————————————————————
Prepared for Members and Committees of Congress

Summary

For most of the twentieth century, the primary use of coal in the United States was for electric power generation, and for most of the history of power generation in the United States, coal has been the dominant fuel used to produce electricity. Even as recently as 2011, coal was the fuel used for almost 42% of power generation in the United States accounting for 93% of coal use. Industrial uses represented the remaining 7%. However, in April 2012, coal's share of the power generation market dropped to about 32% (according to Energy Information Administration statistics), equal to that of natural gas. Coal was the fuel of choice because of its availability and the relatively low cost of producing electricity in large, coal-burning power plants which took advantage of coal's low-priced, high energy content to employ economies of scale in steam-electric production. However, coal use for power generation seems to be on the decline, and the magnitude of coal's role for power generation is in question. Two major reasons are generally seen as being responsible: the expectation of a dramatic rise in natural gas supplies, and the impact of environmental regulations on an aging base of coal-fired power plants.

A recent drop in natural gas prices has been enabled by increasing supplies of natural gas largely due to horizontal drilling and hydraulic fracturing (i.e., fracking) of shale gas formations. If the production can be sustained in an environmentally acceptable manner, then a long-term, relatively inexpensive supply of natural gas could result. Decreased natural gas prices are lowering wholesale electricity prices, stimulating a major switch from coal to gas-burning facilities. The electric utility industry values diversity in fuel choice options since reliance on one fuel or technology can leave electricity producers vulnerable to price and supply volatility. However, an "inverse relationship" may be developing for coal vs. natural gas as a power generation choice based on market economics alone, and policies which allow one fuel source to dominate may come at the detriment of the other.

Coal-fired power plants are among the largest sources of air pollution in the United States. More than half a dozen separate Clean Air Act programs could possibly be used to control emissions, which makes compliance strategy potentially complicated for utilities and difficult for regulators. Because the cost of the most stringent available controls, for the entire industry, could range into the tens of billions of dollars, some power companies have fought hard and rather successfully to limit or delay regulations affecting them, particularly with respect to plants constructed before the Clean Air Act Amendments of 1970 were passed. The expected retirement of approximately 27 GW of coal-fired capacity by 2016 has been reported to the Energy Information Administration (EIA) by coal plant owners and operators, accounting for approximately 8.5% of U.S. coal-fired capacity. While the costs of compliance with new Environmental Protection Agency regulations are a factor, several other issues are cited by coal plant owners and operators as contributing to these retirement decisions including the age of coal-fired power plants, flat to modest electricity demand growth, the availability of previously underutilized natural gas combined-cycle power plants, and the lower price of natural gas due to shale gas development. Even coal plants which have made significant modifications to meet existing EPA regulations are being closed or mothballed due to a combination of low natural gas prices, and the inability to sell power into other markets.

EIA expects coal to be a significant part of the U.S. power generation industry's future to well past 2030. But given price competition from natural gas, and emerging environmental regulations, that role will likely be smaller than in recent decades. Coal-fired generation is likely to face a challenging future.

Contents

Figures

Tables

Contacts

Introduction

Coal has been the fuel most used for thermal energy since the early days of the Industrial Revolution. However, the mechanization of factories quickly brought a shift in coal use from heat to steam engines, and then to fuel steam turbines to power the electric motors of industry. The advent of electrical lighting and the electrification of buildings and residences drove even greater demand for electricity. For most of the twentieth century, the primary use of coal in the United States was for electric power generation.

In 2011, 93% of coal consumed in the United States was for electric power production, according to the Energy Information Administration (EIA). **Figure 1** shows the use of the four major types of coal produced in the United States, with bituminous and subbituminous coal dominating electric power generation.

Figure 1. Flow of U.S. Coal Consumption for 2011
(Million Short Tons)

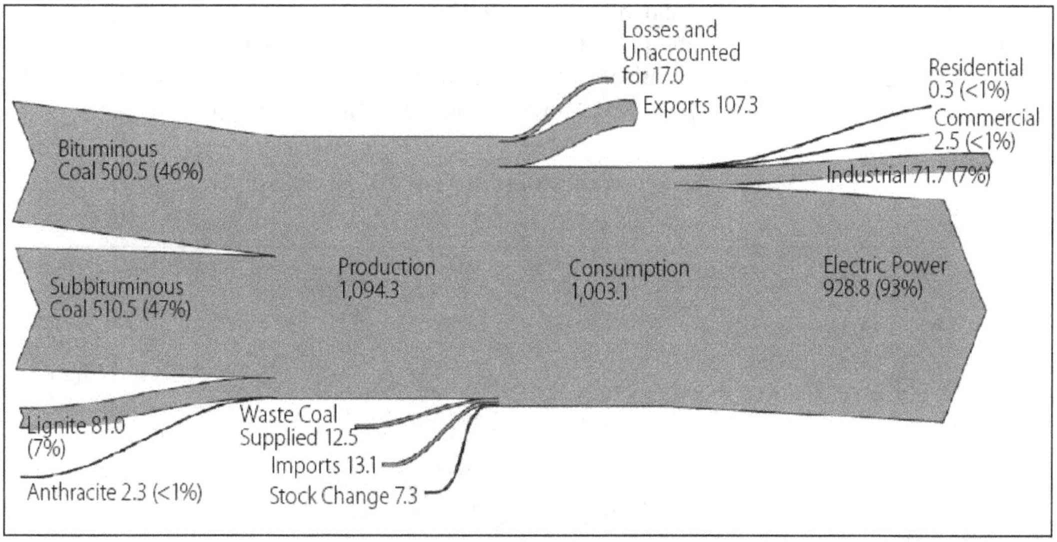

Source: U.S. Energy Information Administration. See http://www.eia.gov/totalenergy/data/annual/pdf/sec7_3.pdf.

The various types of coal differ mostly in carbon content and heating value:

- **Anthracite** has the highest carbon content (between 86% and 98%), and a heat value of about 15,000 British Thermal Units (BTUs) per pound.[1] Anthracite coal is a small part of the electric power market, and is mostly found in the Appalachian region of Pennsylvania.

---

[1] The heating value of any fuel is the energy released per unit mass or per unit volume of the fuel when the fuel is completely burned. Higher heat value fuels liberate more energy per unit of mass or volume.

- **Bituminous** is the most abundant form of coal in the United States, and is the type most commonly used to generate electricity. Bituminous coal has a carbon content of approximately 45% to 86%, and a heat value between 10,500 BTUs and 15,500 BTUs.

- **Subbituminous** coal is mostly found in six western states and Alaska. It has a carbon content of between 35% and 45%, and a heat value of between 8,300 BTUs and 13,000 BTUs. Subbituminous coal generally has a lower sulfur content than other types of coal.

- **Lignite** has the lowest carbon content of the four types of coal generally used for electric power generation, averaging between 25% and 35%, and a high moisture and ash content. It also has the lowest heat value ranging between 4,000 BTUs and 8,300 BTUs.

Electric Power Generation

For most of the history of power generation in the United States, coal has been the dominant fuel used to produce electricity. Even as recently as 2011, coal was used to fuel almost 42% of power generation in the United States, as shown in **Figure 2**. Coal has been the fuel of choice for many decades because of its wide availability, and the relatively low cost of producing electricity in large, coal-burning power plants. Coal's low-priced, high energy content enabled the building of power plants able to take advantage of economies of scale in steam-electric production.

Figure 2. Sources of Electric Power Generation

Net U.S. Generation 2011

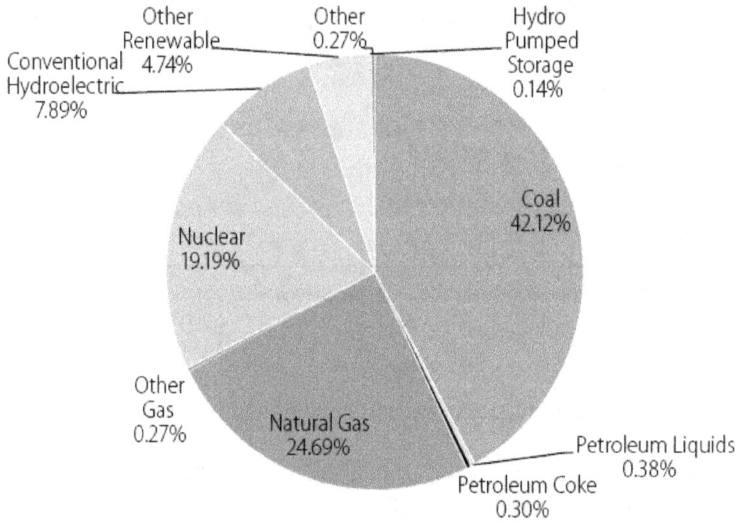

Source: CRS. Data from Energy Information Administration. Electric Power Monthly. Table 1.1, All sources of power.

In a steam power plant, coal (or other fossil fuel) is burned to provide heat for turning water into steam in a boiler.[2] The steam is then forced under pressure into a steam turbine-driven generator

[2] A simple cycle natural gas power plant burns natural gas in a combustion turbine (like a jet engine) to turn a generator
(continued...)

which produces electricity. As of 2011, the U.S. coal-powered generation fleet consisted of 1,387 units with a nameplate capacity of almost 318 gigawatts[3] (GW)[4] representing approximately 30% of total U.S. generating capacity.[5] Electric utilities and independent power producers[6] are the primary entities involved in U.S. electricity production. Together, these electric power sector participants have combined for an average annual consumption of almost 995 million tons of coal in the period from 2002 to 2011,[7] with electric utilities consuming between 60% to 70% of the total (primarily in base load[8] power plants).

The capital costs of power plant construction (i.e., the costs of finance and construction) vary according to the type (for example, coal or nuclear), and size of power plant. The cost for these plants to generate electricity is largely based on their cost of fuel, as well as operations and maintenance costs, other capital costs, and how the power plant is operated (for example, whether the facility is run as a base load or peaking[9] facility).[10] While the regulatory regime (i.e., under a rate-regulated or competitive environment), costs of transmission and distribution, and weather are strong determinants, the cost of fuel is often the major factor affecting electricity production especially for power plants which have already recovered their capital costs of construction.

Transportation costs for coal have historically been a major consideration in the choice of coal as a fuel. Barge transport is used where river access is available. But rail transportation is the most common mode of transit for coal as unit trains[11] offer the ability to transport large shipments to power plants on a regular basis. However, rail transportation can be costly.

> First, coal-burning electric power plants receive approximately 72 percent of their coal by rail. Second, rail transportation costs account for a sizable share of total delivered costs. While, on average, transportation costs account for approximately 20 percent of total [delivered] costs, they can reach as high as 59 percent on shipments of coal originating in the Powder River Basin.[12]

(...continued)

which produces electricity. When used in a combined cycle mode, the hot exhaust gases from the combustion turbine are used to generate steam to turn a steam turbine to efficiently create additional electricity.

[3] One billion watts or one thousand megawatts.

[4] See http://www.eia.gov/todayinenergy/detail.cfm?id=7290.

[5] See http://www.eia.gov/totalenergy/data/annual/showtext.cfm?t=ptb0811a.

[6] A corporation, person, agency, authority, or other legal entity or instrumentality that owns or operates facilities for the generation of electricity for use primarily by the public, and that is not an electric utility. See http://www.eia.gov/tools/glossary/index.cfm?id=I#ind_pwr_prod.

[7] EIA, *Table 2.1.A. Coal: Consumption for Electricity Generation*, 2002-2011, http://www.eia.gov/electricity/monthly/pdf/epm.pdf.

[8] A plant, usually housing high-efficiency steam-electric units, which is normally operated to take all or part of the minimum load of a system, and which consequently produces electricity at an essentially constant rate and runs continuously. These units are operated to maximize system mechanical and thermal efficiency and minimize system operating costs. EIA, *EIA Glossary "Base Load Plant"*, http://www.eia.gov/tools/glossary/index.cfm.

[9] A plant usually housing old, low-efficiency steam units, gas turbines, diesels, or pumped-storage hydroelectric equipment normally used during the peak-load periods. See http://www.eia.gov/tools/glossary/index.cfm.

[10] Black and Veatch, *Cost Report—Cost and Performance Data for Power Generation Technologies*, National Renewable Energy Laboratory, February 2012, http://bv.com/docs/reports-studies/nrel-cost-report.pdf.

[11] Typically, all the freight cars on a "unit train" are filled with coal being shipped directly to a single destination.

[12] See http://www.eia.gov/coal/transportationrates/.

U.S. coal transportation costs are cited as a major reason why many utilities in the New England states have chosen to generate electricity with natural gas, or import coal from overseas as a lower cost alternative.[13]

Recent Changes in Markets for Electricity Fuels

The amounts of coal consumed for electricity generation from 2002 to 2011 are shown in **Figure 3**. While coal was the dominant fuel for electricity production in the period, the trend of increasing consumption seems to be slowing with a recent downturn in coal consumption for electricity beginning in about 2008. At the same time, natural gas use for electric power production is rising.

Figure 3. Fuel Consumption for Electricity Generation 2002-2011

Electric Utilities and Independent Power Producers

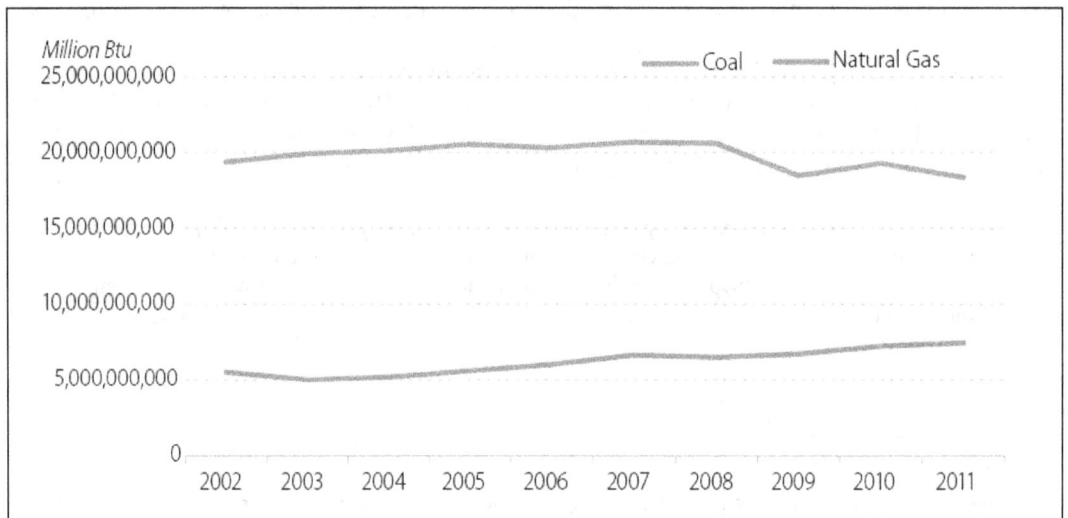

Source: CRS, data from U.S. Energy Information Administration (EIA), *Electric Power Monthly*, Table 2.1.A, "Coal: Consumption for Electricity Generation by Sector," at http://www.eia.gov/electricity/monthly/pdf/epm.pdf, September 2012; CRS, data from U.S. Energy Information Administration (EIA), *Electric Power Monthly*, Table 2.4.A, "Natural Gas: Consumption for Electricity Generation by Sector," at http://www.eia.gov/electricity/monthly/pdf/epm.pdf.

Increasing supplies of natural gas are largely behind the switch in fuel for power generation. Natural gas prices have been volatile over the last few decades, ranging from less than $2 per million Btus (MMBTU) to over $18 per MMBTU.[14] However, supplies of natural gas increased markedly beginning in 2008-2009 largely due to development of unconventional shale gas through improved drilling and simulation techniques. Horizontal drilling and hydraulic fracturing

---

[13] See ISO New England, "Scenario Analysis Project—Long Term Forecast of Oil, Natural Gas and Coal Prices in New England," 2007, http://www.isone.com/committees/comm_wkgrps/othr/sas/mtrls/apr22007/fuel-price-forecast.pdf.

[14] See Energy Information Administration, Office of Oil and Gas, *An Analysis of Price Volatility in Natural Gas Markets*, "Figure 1: Daily Henry Hub Spot Price, 1994-2006," August 2007, http://www.eia.gov/pub/oil_gas/natural_gas/feature_articles/2007/ngprivolatility/ngprivolatility.pdf.

(i.e., fracking) have dramatically improved shale gas production, and the increased supplies of natural gas have caused prices to fall.

Power Plant Dispatch and Competitive Power Production

Many regions of the United States use competitive markets for power generation which sell into the wholesale market. These markets are generally administered by regional transmission organizations (RTOs) which have a goal of minimizing the cost of wholesale electricity to distribution utilities. The distribution utilities sell power to end-use (retail) consumers. In RTO markets, power plants bid against each other to serve the electricity demand of the wholesale market. Operators "dispatch" (or schedule the operation of) power plants based on the least costly choice, considering forecast loads and transmission system constraints for specific load increments at various times of the day or year. The most efficient power plants with the lowest heat rates (the efficiency with which a power plant turns fuel into electricity) are typically dispatched more, and lower efficiency, higher heat rate units are dispatched less.

Lower natural gas prices have caused wholesale electricity prices to decline, lowering potential revenues. However, many of the newer, more efficient natural gas combined cycle power generation plants have heat rates around 7,000 BTU/kWh (compared to today's coal plants with an overall average heat rate of 10,000 BTU/kWh). As natural gas prices have declined, natural gas plants have been dispatched more, and gross margins have increased with the spark spread (a measure of the potential profitability of a natural gas-fired power plant based on the difference between the price that a generator can obtain from selling one megawatt hour (MWh) of electricity and the cost of the natural gas needed to generate the MWh of electricity).

For coal plants, the lower cost of wholesale power has combined with the rising costs of coal to decrease their potential profitability. Older, less efficient coal power plants are especially at risk because wholesale prices of electricity may be inadequate for such facilities to operate economically.

Figure 4 shows natural gas and coal fuel costs for electricity production from 2002 to 2012. The right-most part of the chart is based on monthly data, illustrating the recent drop in natural gas prices for electricity production while coal costs continue to inch higher. The drop in natural gas prices is a major factor in the recent switch from coal to gas-burning facilities, since the cost of fuel is a major determinant in the dispatch of electricity production. Switching from coal to gas generation has largely involved a switch from base load coal generation to intermediate load[15] natural gas combined cycle (NGCC) capacity. There are varying amounts of underutilized NGCC capacity available in various regions of the United States. That underutilized capacity resulted largely from the natural gas "bubble" of the 1980s and 1990s when natural gas prices were generally about $2 per MMBTU.

Fuel cost to generate electricity is the key in the decision to switch from coal to natural gas generation. If a power plant was 100% efficient, it would take 3.412 MMBTU to generate 1 MWh of electricity. Assuming that an intermediate capacity NGCC plant is 44.78% efficient, it would consume 7.619 MMBTUs to generate 1 MWh of electricity. Similarly, a base load coal plant with an efficiency of 33.64% would use 10.412 MMBTUs to generate 1 MWh of electricity.[16] The cost of fuel to generate 1 MWh of electricity from coal has historically been lower than the cost to generate power from natural gas. The economics are now changing with coal prices rising and the drop in natural gas prices in 2011 down to levels where it now becomes

---

[15] The range from base load to a point between base load and peak load.

[16] Sandy Fielden, "The Economics of Coal-to-Gas Switching," *Power Magazine*, August 21, 2012, http://www.powermag.com/business/The-Economics-of-Coal-to-Gas-Switching_4880_p4.html.

---

more expensive to generate electricity from coal, especially when power plant efficiency is taken into account.

Figure 4. Cost of Fuel for Electricity Generation, 2002-2011

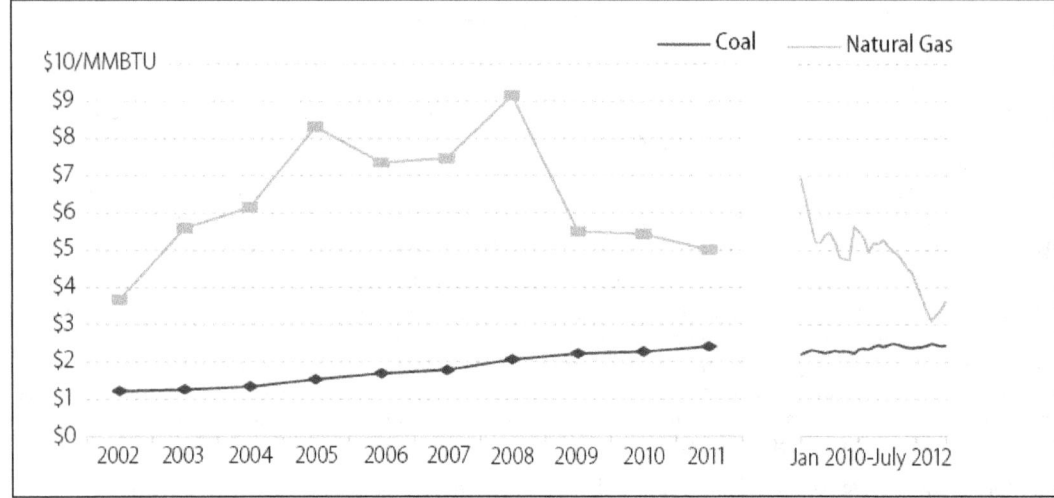

Source: CRS, data from EIA, "Table 4.2, Receipts, Average Cost, and Quality of Fuel, Electric Utilities" at http://www.eia.gov/electricity/monthly/pdf/epm.pdf. CRS. Data from EIA, Table 4.2, Receipts, Average Cost, and Quality of Fuel, Electric Utilities at http://www.eia.gov/electricity/monthly/pdf/epm.pdf.

While the chart in **Figure 5** shows the expected rise and fall of power production with the seasons in the 2009 to 2012 timeframe, a look behind the numbers reveals the shift in generation as more power is being dispatched from higher efficiency natural gas-fired power plants.[17] During this time, coal's share of total U.S electricity generation has dropped and the use of natural gas to generate electricity has risen as newer, more efficient natural gas-fired generating capacity are being dispatched.[18] In April 2012, for the first time in history, the amount of electricity generation from natural gas equaled that of coal, according to EIA statistics, each with about 32% of the market.[19]

---

[17] EIA, "Natural Gas Spot Prices near 10-Year Lows Amid Warm Weather and Robust Supplies," February 1, 2012, http://www.eia.gov/todayinenergy/detail.cfm?id=4810.

[18] "Coal generation decreased 29 billion kilowatt-hours from March 2011 to March 2012, while natural gas generation increased 27 billion kilowatt-hours during the same time period. In March 2012, coal's share of total generation was 34% compared to natural gas at 30%." See EIA, "U.S. Coal's Share of Total Net Generation Continues to Decline," June 5, 2012, http://www.eia.gov/todayinenergy/detail.cfm?id=6550.

[19] U.S. Energy Information Administration, *Net Generation by Energy Source: Total (All Sectors), 2002-June 2012*, July 2012, http://www.eia.gov/electricity/monthly/epm_table_grapher.cfm?t=epmt_1_1.

Figure 5. Net Generation—Coal vs. Natural Gas

Period from 2009 to 2012

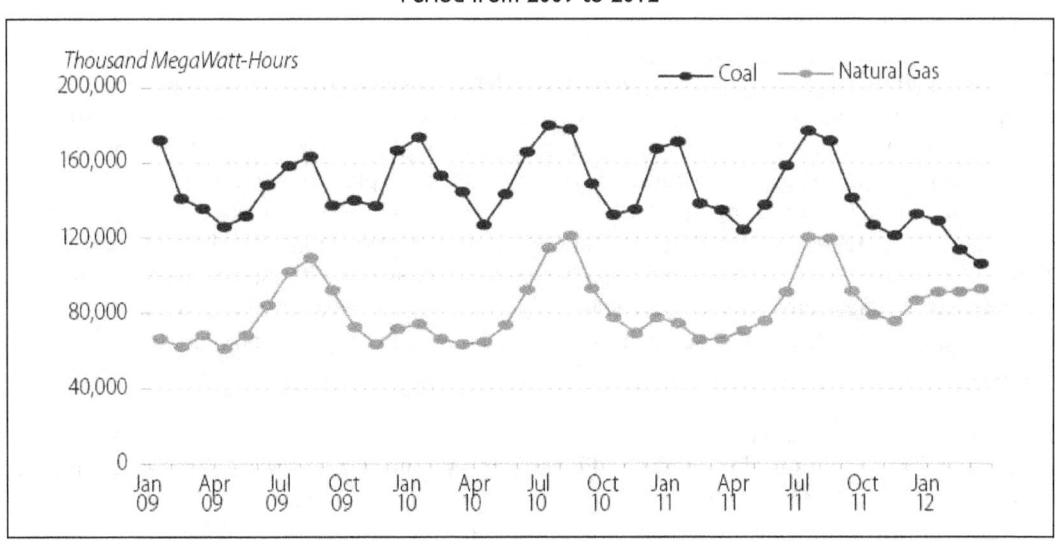

Source: CRS, data from EIA. See http://www.eia.gov/todayinenergy/detail.cfm?id=6550.

Many in the energy industry believe that a structural change in the economics of natural gas use has begun.

> While coal is projected to retain the largest share of the electricity generation mix through 2035, analyses included in the Annual Energy Outlook 2012 ... anticipate its share declining as more generation comes from natural gas and renewable technologies. Coal's role as the preeminent source of electricity generation in the United States has lessened in recent years, declining from 49% of total electricity generation in 2007 to 42% in 2011.... Projected fuel prices and economic growth are key factors influencing the future electricity generation mix. The price of natural gas, coal's chief competitor, has dropped significantly in recent years due to the increase in domestic production of natural gas.[20]

Nevertheless, there is still much debate as to whether the "shale gas revolution" will result in a predictable, long-term increase in supplies of natural gas at relatively low prices.

> Considerable uncertainty exists regarding the size of the economically recoverable U.S. shale gas resource base and the cost of producing those resources. Across four shale gas resource scenarios from the [EIA's] Annual Energy Outlook 2012, natural gas prices vary by about $4/MMBtu in 2035, demonstrating the significant impact that shale gas resource uncertainty has in determining future natural gas prices. This uncertainty exists primarily because shale gas wells exhibit a wide variation in their initial production rate, rate of decline, and estimated ultimate recovery per well (or EUR, which is the expected cumulative production over the life of a well).[21]

---

[20] EIA, "Fuel Used in Electricity Generation Is Projected to Shift over the Next 25 Years," July 30, 2012, http://www.eia.gov/todayinenergy/detail.cfm?id=7310.

[21] EIA, Projected Natural Gas Prices Depend on Shale Gas Resource Economics," August 27, 2012, http://www.eia.gov/todayinenergy/detail.cfm?id=7710.

If these previously unconventional shale gas resources can be economically developed and produced in an environmentally acceptable manner, then a sustained, relatively inexpensive supply of natural gas could persist.[22]

As long as natural gas prices are "relatively" high, development of resources will continue. When a surplus of natural gas supplies causes the price to drop, resource development activities solely focused on natural gas tend to curtail. But fracking of shale gas resources is changing the outlook for natural gas supplies, and the recovery of oil from wet shale plays is currently driving development. Even so, the expectation of lower, less volatile prices from increased supplies is encouraging the greater use of natural gas for electric power generation and other industrial uses. Natural gas prices are still subject to forces of supply and demand, with the cost of developing incremental supplies determining future supplies of the resource. Further development of the interstate pipeline system may also be needed if these new natural gas supplies are to reach markets.[23]

How the natural gas industry will accommodate the expected demands of the power generation industry for gas supplies is also uncertain. Power plants purchase coal under long-term contracts, amassing the fuel on-site in large, managed coal storage piles which may hold a 30-day supply. Most natural gas power generation plants are served directly by pipelines, and manage the fuel as more of a "just-in-time" resource, since power plants typically do not have large natural gas storage capabilities. The Federal Energy Regulatory Commission observes that based on recent supply issues, this new volume of expected demand for natural gas may require more longer term coordination and planning between the electricity and natural gas supply industries.[24]

Potential Coal Plant Retirements

All power plants are subject to retirement when they reach they reach the end of their useful service life. While older coal-fired power plants are usually well-maintained, they are generally not as efficient as newer power plants. As power plants age, they are generally upgraded to continue operations, but the least efficient plants may be retired. Other plants may be shifted from base load operations (in which they essentially operate around the clock) to less demanding intermediate or peaking schedules. The cost of building a power plant is generally recovered over the depreciable life of the asset, such that operations and maintenance (O&M) expenses become the major component of an older power plant's continuing costs. A major component of O&M is the cost of fuel, and the expectation of continued lower prices for natural gas is weighing on decisions concerning whether many older, less efficient coal power plants will be mothballed or closed altogether. However, the costs of modernizing older power plants to meet new regulatory requirements can be relatively high. When the cost of upgrades to meet new environmental requirements is considered along with (perhaps increasing) O&M expenses, many older coal power plants are likely to face outright retirement decisions.

---

[22] For further discussion of natural gas production, see CRS Report R42814, *Natural Gas in the U.S. Economy: Opportunities for Growth*, by Robert Pirog and Michael Ratner

[23] Jack Buehrer, *Natural Gas Boom Drives Pipeline Upgrades*, Engineering News Record, May 7, 2012, http://enr.construction.com/infrastructure/power_industrial/2012/0507-shale-boomhttpprimisphmsadotgovconstructionconstphmsahtm.asp.

[24] Federal Energy Regulatory Commission, *Natural Gas - Electric Coordination*, August 2012, http://www.ferc.gov/industries/electric/indus-act/electric-coord.asp.

Over the last 40 years, Congress has directed EPA to reduce the potential health and environmental impacts of fossil fuel use by limiting emissions by-products or other consequences of combustion processes.[25] These environmental regulatory requirements have been evolving in the last decade due to various challenges to the regulatory implementation of federal laws. Much attention has focused recently on the resulting finalization of some of these regulations, and their potential to contribute to the retirement of some coal-burning power plants.

In addition to being the largest source of electric power, coal-fired power plants are among the largest sources of air pollution in the United States. Under the Clean Air Act (CAA),[26] however, they have not necessarily been subject to stringent requirements: emissions and the required control equipment can vary depending on the location of the plant, when it was constructed, whether it has undergone major modifications, the specific type of fuel it burns, and, to some extent, the vagaries of EPA enforcement policies. More than half a dozen separate CAA programs could potentially be used to control emissions, which makes compliance strategy potentially complicated for utilities and difficult for regulators. Because the cost of the most stringent available controls, for the entire industry, could range into the tens of billions of dollars, power companies have fought hard and rather successfully to limit or delay regulations affecting them, particularly with respect to plants constructed before the Clean Air Act of 1970 was passed. However, EPA contends that such costs would be more than outweighed by societal benefits from reduced respiratory and other illnesses—"externalities" that are largely not currently accounted for in the price of electricity generation but which are estimated by EPA as costing the United States billions of dollars annually in health care and environmental costs.[27]

Some of the new rules under development at the EPA would be implemented at the federal level, while others would be implemented at the state level. They include the Cross-State Air Pollution Rule[28] (which replaced the Clean Air Interstate Rule); the Utility Maximum Achievable Control Technology (Utility MACT) rule to reduce emissions of mercury, other metallic toxics, acid gases, and organic air toxics; the wet ash classification rule for coal combustion residues; and the Clean Water Act Section 316(b) guidelines for once-through cooling water systems. However, only the Utility MACT rule is currently in effect. EPA also proposed standards for greenhouse gas

---

[25] For example, with the Clean Air Act Amendments of 1970 (P.L. 91-604) and subsequent revisions, the Clean Water Act of 1972 (P.L.92-500), and the Resource Conservation and Recovery Act of 1976 (P.L. 94-580).

[26] The Clean Air Act, codified as 42 U.S.C. 7401 et seq., seeks to protect human health and the environment from emissions that pollute ambient, or outdoor, air. It requires the Environmental Protection Agency to establish minimum national standards for air quality, and assigns primary responsibility to the states to assure compliance with the standards. Areas not meeting the standards, referred to as "nonattainment areas," are required to implement specified air pollution control measures. The Act establishes federal standards for stationary and mobile sources of air pollution and their fuels and for sources of 187 hazardous air pollutants, and it establishes a cap-and-trade program for the emissions that cause acid rain. It establishes a comprehensive permit system for all major sources of air pollution. It also addresses the prevention of pollution in areas with clean air and protection of the stratospheric ozone layer. For more information, see CRS Report RL30853, *Clean Air Act: A Summary of the Act and Its Major Requirements*, by James E. McCarthy, Claudia Copeland, and Linda-Jo Schierow.

[27] USEPA Office of Air and Radiation, *The Benefits and Costs of the Clean Air Act: 1990 to 2020*, August 2010, http://www.epa.gov/air/sect812/aug10/fullreport.pdf.

[28] On August 21, 2012, in a 2-1 decision, the court vacated and remanded the rule, finding that EPA's imposition of Federal Implementation Plans, without first giving the states an opportunity to develop their own plans, was unlawful. See CRS Report R41563, *Clean Air Issues in the 112th Congress*, by James E. McCarthy. By vacating the rule, the Court determined that the Clean Air Interstate Rule would still apply. But that rule was itself vacated, making future implementation of any air transport rule unclear.

(GHG) emissions which would require all new power plants to restrict carbon dioxide emissions.[29]

EPA has yet to propose rules for GHG emissions from existing power plants, as is required by court order. Due to a general perception in the electric power industry that these new and pending environmental regulations present conflating requirements with unrealistic timeframes for compliance, the regulations have come to be referred to by the industry as the "train wreck" scenario[30] due to a perception that a negative impact on reliability could result. However, environmental groups, and some in the electric power industry—mainly those with significant investments in nuclear or natural gas-fired generation—consider the concerns overstated.[31]

Electricity Reliability—State and Market Inputs

As retirement decisions are announced for growing amounts of coal capacity, questions are being asked about the cost and types of generation which may replace that capacity. EIA defines electric system reliability as "the degree to which the performance of the elements of the electrical system results in power being delivered to consumers within accepted standards and in the amount desired." Implicit in the definition is the necessity of generation adequacy to ensure that there is sufficient power plant capacity to meet projected demand, and additional reserves for exceptionally high demand periods. This includes voltage support (i.e., power injected into the grid to maintain voltage at acceptable limits) and ancillary services (i.e., services which support the reliable operation of the transmission system as it moves electricity from generating sources to retail customers) provided by retiring plants. Timely and appropriate replacement of retiring coal capacity would be essential to maintain reliability levels.

Cost recovery for a new power plant entails not only financial and performance risks, but also potential regulatory risk. In traditional cost of service states, adding a new power plant to a utility's rate base can raise customer rates. While alternatives to potentially expensive new power plant construction may exist (such as demand response), if it appears that adding a new plant or plants to rate base will result in a significant rate increase, a reluctance may ensue on the part of state officials to authorize the addition especially in times of economic hardship. Similarly, investor-owned utilities are wary of taking on too much debt for new power plant construction as this may decrease earnings and share price, possibly resulting in a downgrade of a company's financial rating.

In states with traditional cost of service ratemaking (in which electric utility rates are approved by state utility commissions), formal authorization is required before electric utilities can site, build, or place power plants into service thereby receiving recovery of costs in electricity rates. The need for new power plants is typically determined years in advance by a formal process estimating future demand and the ability of existing generation resources to meet that demand. A commonly used process is integrated resource planning (IRP), which considers a variety of supply and demand-side options, and assesses them against a common set of planning objectives and criteria, including environmental impact. The IRP objective is to help meet future customer demand by identifying the need for generating capacity, and determining the best mix of resources (including demand response) to fill the need at the lowest overall cost. Electric utilities often seek to shift the financial risk of building new power plants to ratepayers by obtaining regulatory approval to begin collecting the cost of the power plant in customer rates before the plant is in service (i.e., as construction work-in-progress).

Even in RTO markets where power generation is a competitive business, siting of new power plants is subject to state regulatory processes. However, in competitive markets, the financial risk of building a new power plant lies with the generator as they seek to bid to sell electricity into the RTO market. RTOs use markets to make operational decisions (such as generator dispatch) and to determine the provider(s) and prices for many of these services. RTOs

---

[29] Environmental Protection Agency, *Carbon Pollution Standard for New Power Plants*, http://epa.gov/carbonpollutionstandard/.

[30] These regulations are discussed in separate CRS Reports: CRS Report R41914, *EPA's Regulation of Coal-Fired Power: Is a "Train Wreck" Coming?*, by James E. McCarthy and Claudia Copeland, and CRS Report R42570, *Proposals to Amend RCRA: Analysis of Pending Legislation Applicable to the Management of Coal Combustion Residuals*, by Linda Luther.

[31] CRS Report R42144, *EPA's Utility MACT: Will the Lights Go Out?*, by James E. McCarthy.

---

do not own generation, but rely on price signals (based on real-time prices) to incentivize entry of new generation resources. Some RTOs employ forward capacity markets (up to three years in advance of expected needs) with auctions using either vertical demand curves (establishing quantity and allowing price to be set by the market) or sloped demand curves (regulating the price of new capacity) to contract for future capacity. Other RTOs use an administrative approach to determine capacity requirements and set prices. However, there is some debate as to how these market approaches work, and the resulting impact on prices. Other issues include whether forward capacity markets result in new capacity actually being built, whether the appropriate type of capacity (i.e., base load vs. intermediate load) is being incentivized and attracted, or even if the administratively-set or auction-determined capacity price is above the actual cost of new entry in these markets. The issue of prevailing high electricity prices in some areas (i.e., in load pockets) has led some states in RTO regions to consider authorizing the construction of power plants or bilateral contracting for electric power independent of RTO capacity market processes.

In 2011, the North American Reliability Corporation (which has oversight of the bulk electric grid for reliability purposes) analyzed the effects of the proposed EPA regulations for their potential to accelerate retirement of fossil-fired units.[32] The 2011 assessment was a revision of a previous analysis to take into account "updated assumptions determined from proposed rules, as well as other viewpoints on how the rule will ultimately be carried out (i.e., how states may implement regulations), to determine potential resource adequacy impacts." NERC noted that the expected retirements of electric generators are driven by many economic factors, "not simply the cost of pollution control equipment." However, NERC's assessment estimated that between 35 GW and 59 GW of generating capacity could be potentially retired as a result of the proposed EPA rules. NERC summarized its concerns as to the possible effects on reliability:

> Existing and proposed environmental regulations in the U.S. may significantly affect bulk power system reliability depending on the scope and timing of the rule implementation and the mechanisms in place to preserve reliability.[33]

For its part, the Department of Energy (DOE) estimated in 2011 that new air pollution, water, and waste disposal proposed rules could eventually result in the retirement of between 35 GW and 70 GW of coal-fired capacity.[34]

The expected retirement of approximately 27 GW of coal-fired capacity by 2016 has been reported to EIA by coal plant owners and operators, accounting for approximately 8.5% of U.S. coal-fired capacity. While the costs of compliance with new EPA environmental regulations are a factor, several other issues are cited by coal plant owners and operators as contributing to these retirement decisions including the age of coal-fired power plants, modest electricity demand growth, the availability of previously underutilized natural gas combined-cycle power plants,[35]

---

[32] North American Electric Reliability Corporation, *Potential Impacts of Future Environmental Regulations—Extracted from the 2011 Long-Term Reliability Assessment*, November 2011, http://www.nerc.com/files/EPA%20Section.pdf.

[33] Ibid.

[34] Debra McCown, "EPA Regulations for Coal-Fired Power Plants Could Force Shut Downs," *Bristol Herald Courier*, May 24, 2011, http://www2.tricities.com/business/2011/may/25/wood-gives-dire-warning-due-epa-regulations-coal-f-ar-1062322/.

[35] An electric generating technology in which electricity is produced from otherwise lost waste heat exiting from one or more gas (combustion) turbines. The exiting heat is routed to a conventional boiler or to a heat recovery steam generator for utilization by a steam turbine in the production of electricity. This process increases the efficiency of the electric generating unit. See http://www.eia.gov/tools/glossary/index.cfm.

and the lower price of natural gas due to shale gas development.[36] And some coal plants which have made significant modifications to meet existing EPA regulations are being closed or mothballed due to a combination of low natural gas prices, and either transmission congestion[37] or a lack of transmission options to sell power into other markets.[38]

Figure 6. Percentage of Electricity Generated from Coal by State, 2010

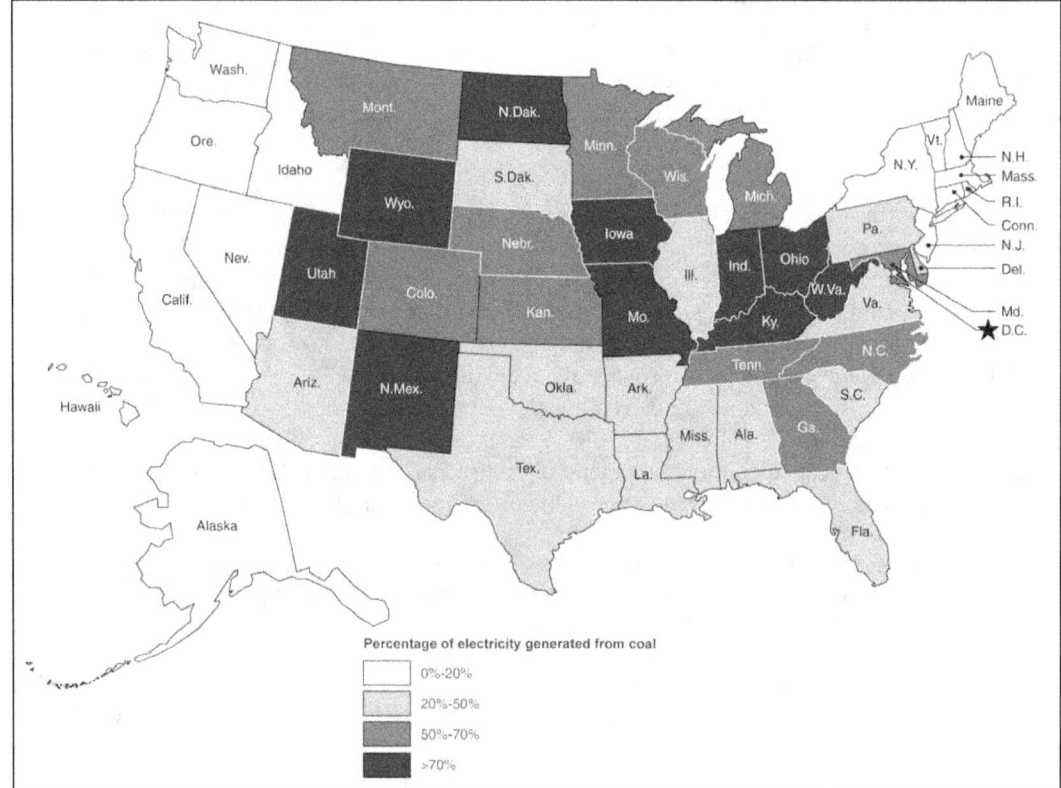

Source: Government Accountability Office. See http://www.gao.gov/assets/600/592542.pdf.

The implications of coal-fired electric power plant retirements will likely play out on a regional basis, as EPA's rules go through inevitable challenges and modifications as the regulations are finalized.[39] Power generation in the United States varies according to local energy resources, and cost and availability of infrastructure for fuels transportation.[40] The Eastern (except New

---

[36] EIA, *27 GigaWatts of Coal-Fired Capacity to Retire over Next Five Years*, July 27, 2012, http://www.eia.gov/todayinenergy/detail.cfm?id=7290#.

[37] A condition that occurs when insufficient transfer capacity is available to implement all of the preferred schedules for electricity transmission simultaneously. See http://www.eia.gov/tools/glossary/index.cfm.

[38] Gib Snyder, "An Inside Look at the Process for Mothballing," *The Observer*, September 9, 2012, http://www.observertoday.com/page/content.detail/id/575844/An-inside-look-at-the-process-for-mothballing.html?nav=5047.

[39] InsideEPA, *EPA Urges Appellate Court To Rehear Ruling Vacating Cross-State Air Rule*, October 5, 2012, http://insideepa.com/201210052412215/EPA-Daily-News/Daily-News/epa-urges-appellate-court-to-rehear-ruling-vacating-cross-state-air-rule/menu-id-95.html.

[40] "In New England, most of the coal that will be burned ... will likely originate in South America or Indonesia, where the chemical composition of the coal is less harmful to the environment. Western U.S. coals also have a favorable (continued...)

England) and Midwest regions of the United States are more coal-dependent than the Northeast or Western regions, as shown in **Figure 6**.

Several large regional transmission organizations[41] (RTOs) have undertaken studies of the impact of the new EPA rules. A major question being asked in these studies is whether there will be a risk to reliability based on the timing of requirements, and the scheduling and availability of equipment and personnel to accomplish retrofits of power plants to bring them into compliance with EPA regulations. The results of a study[42] by the Midwest ISO (MISO), the RTO serving 11 states and one Canadian province may provide insights into the implications of both EPA regulations and lower natural gas prices. There is approximately 70 GW of coal-fired capacity in the MISO market,[43] with an average coal plant age of 45 years.[44] The study identified potential retirements because of proposed EPA rules, a low natural gas price, and various levels of carbon mitigation costs. MISO summarized the results of its study noting that EPA regulations and lower natural gas prices would likely result in between 3 GW and 13 GW of coal capacity retirements, with MISO itself likely to see costs of between $32 billion and $33 billion due to EPA compliance retrofit costs, new capacity charges, system energy costs, and transmission system upgrades to maintain reliability.[45]

MISO does not own or build power plants. State utility commissions still control the process of new power plant construction in many states. States often require a utility to use integrated resource planning methods[46] to address system needs. If a determination is made that new generating capacity is needed to ensure reliability, then how to requisition that capacity becomes the responsibility of state public utility commissions or the RTO markets. The decision would likely be made according to how they view the effects of coal plant retirements on the levels and types (i.e., base load vs. intermediate or peak load power plants) of capacity that will be needed to serve electricity demands.[47] RTO markets generally rely on price signals[48] to provide an incentive

---

(...continued)

chemical composition, but it is currently too expensive to deliver coal from the Rocky Mountains to New England relative to foreign supplies." ISO New England, *Scenario Analysis Project—Long Term Forecast of Oil, Natural Gas and Coal Prices In New England*, 2007, http://www.isone.com/committees/comm_wkgrps/othr/sas/mtrls/apr22007/ fuel-price-forecast.pdf.

[41] The purpose of a Regional Transmission Organization is to ensure the efficient and reliable delivery of power across large areas. RTOs facilitate and promote efficiency in wholesale electricity markets and ensure that electricity consumers pay the lowest price possible for reliable service by removing transmission barriers between buyers and sellers. See http://www.pjm.com/about-pjm/learning-center/pjm-overview/energy-basics/what-is-a-rto.aspx.

[42] MISO, *EPA Impact Analysis- Impacts from the EPA Regulations on MISO*, October 2011, https://www.midwestiso.org/Library/Repository/Study/MISO%20EPA%20Impact%20Analysis.pdf. (MISO EPA)

[43] Of the 70,000 MW of coal-fired capacity in the MISO market, less than half does not have plans for SO_2 controls. Furthermore, 38% have no SO_2 controls or NOx controls, and 38% have no SO_2 controls or Fabric Filters. (MISO EPA)

[44] The life expectancy of coal-fired power plants from a cost recovery basis is generally estimated to be 40 years.

[45] MISO EPA.

[46] "Integrated resource planning [IRP] is a crucial element for success in a constantly changing business and regulatory environment and is based on comprehensive, holistic and risk aware analysis. The integrated approach considers a broad spectrum of feasible supply and demand-side options and assesses them against a common set of planning objectives and criteria, including environmental impact. The IRP objective is to help meet future customer demand by identifying the need for generating capacity and determining the best mix of resources to fill the need. The capacity gap is the difference between the projected firm (or known) requirements and existing firm supply." See TVA, *TVA's Environmental and Energy Future*, Integrated Resource Plan, March 2011, p. 27, http://www.tva.gov/environment/ reports/irp/pdf/Final_IRP_Ch1.pdf.

[47] Many electric utilities are required by states to undertake long-range planning. "Integrated resource planning (continued...)

---

for new generation to enter the market, or employ forward capacity markets with higher prices to encourage new generation to be built. Other "generation-like" resources (such as demand response,[49] energy efficiency, and conservation) would also likely be employed to reduce the need for building new capacity.

Given that significant underutilized NGCC exists in various U.S. regions, the possibility of further shifting from coal base load plants to natural gas intermediate capacity exists. A recent study[50] by the Massachusetts Institute of Technology in 2011 noted that the existing U.S. NGCC generation fleet had an average capacity factor[51] of approximately 41%, while its design capacity allowed such plants to operate at 85%. The MIT study looked at a scenario across selected regions of the United States which mimicked the "full dispatch" of existing natural gas combined cycle plants. The study concluded that under such a scenario (while noting that transmission constraints exist), there is "sufficient surplus NGCC capacity to displace roughly one-third of U.S. coal generation, reducing CO2 emissions from the power sector by 20%."[52]

Role of New and Alternative Power Generation Technologies

According to the EIA, coal is expected to be a key part of electricity generation in the United States well past the year 2030.[53] Electric utilities value diversity in power generation options since reliance on one fuel or technology can leave them vulnerable to price and supply volatility. However, the choices for new power plant technologies will likely be subject to proposed EPA rules limiting GHG emissions. The proposed rule published under the Clean Air Act's New Source Performance Standards program would limit emissions from new power plants to no more than 1,000 pounds of carbon dioxide for each megawatt-hour (MWh) of electricity produced. Today's NGCC generation emits on average about 800 pounds of carbon per MWh compared to current U.S. coal plants with average emissions of about 1,800 pounds of carbon dioxide per each MWh of electric power produced.[54] The proposed GHG rule would allow an exception for the

---

(...continued)

helps electric utilities choose the best resource options to generate electricity and other options to meet customer expectations for energy services. Increasing competition, changing technologies, and environmental concerns are among the many issues utilities must consider when developing their plans." See "Purpose of and Need for Integrated Resource Planning" at http://www.tva.gov/environment/reports/energyvision2020/ev2020_vol1ch01.pdf.

[48] Shortages of capacity cause electricity prices to rise in a competitive market.

[49] Demand response (also known as load response) is end-use customers reducing their use of electricity in response to power grid needs, economic signals from a competitive wholesale market or special retail rates. See http://www.pjm.com/markets-and-operations/demand-response.aspx.

[50] Massachusetts Institute of Technology, *The Future of Natural Gas*, 2011, http://mitei.mit.edu/system/files/ NaturalGas_Report.pdf. (MIT Study)

[51] The ratio of the electrical energy produced by a generating unit for the period of time considered to the electrical energy that could have been produced at continuous full power operation during the same period. See http://www.eia.gov/tools/glossary/index.cfm.

[52] MIT Study, op. cit.

[53] "In the Reference case, the natural gas share of electric power generation increases from 24 percent in 2010 to 28 percent in 2035, while the renewables share grows from 10 percent to 15 percent. In contrast, the share of generation from coal-fired power plants declines. The historical reliance on coal-fired power plants in the U.S. electric power sector has begun to wane in recent years. Over the next 25 years, the share of electricity generation from coal falls to 38 percent, well below the 48-percent share seen as recently as 2008, due to slow growth in electricity demand, increased competition from natural gas and renewable generation, and the need to comply with new environmental regulations." See http://www.eia.gov/forecasts/aeo/pdf/0383(2012).pdf.

[54] 40 C.F.R. § Part 60. See http://epa.gov/carbonpollutionstandard/pdfs/20120327proposal.pdf.

---

approximately one dozen new coal plants that are permitted and beginning construction within one year, and would not apply to existing plants. Only new coal plants with carbon capture and storage (CCS) are likely to meet the proposed GHG regulations,[55] and CCS technology is estimated to add about 30% to a coal plant's cost. However, according to EIA, there are no new conventional coal plants expected to come on line after 2012; 2 GW of coal plant capacity with CCS is expected to enter service by about 2017.[56] New advanced NGCC plants entering the market in 2017 are expected to cost approximately $63 per MWh, while new advanced coal plants are estimated to cost $111 per MWh, and advanced coal with CCS is estimated at $139 per MWh.[57] Based on levelized cost estimates alone, new coal generation (even without CCS) may not be competitive with new NGCC given the built-in expectation of future fuel costs in the estimates.

Carbon Capture and Sequestration (CCS)[58]

Carbon capture and sequestration (or storage)—known as CCS—has attracted congressional interest as a measure for mitigating global climate change. Large amounts of carbon dioxide (CO_2) emitted from fossil fuel use in the United States are potentially available to be captured and stored underground using CCS, and prevented from reaching the atmosphere. Large, industrial sources of CO_2, such as coal-fired electricity-generating plants, are likely initial candidates for CCS because they are predominantly stationary, single-point sources. Electricity generation contributes over 40% of U.S. CO_2 emissions from fossil fuels. Currently, U.S. power plants do not capture large volumes of CO_2 for CCS.

To date, there are no commercial ventures in the United States that capture, transport, and inject industrial-scale quantities of CO_2 solely for the purposes of carbon sequestration. In 2012, the U.S. Department of Energy's CCS research, development, and deployment program embarked on commercial-scale demonstration projects for CO_2 capture, injection, and storage. The success or failure of these projects will likely bear heavily on the future outlook for widespread deployment of CCS technologies as a strategy for preventing large quantities of CO_2 from reaching the atmosphere while U.S. power plants continue to burn fossil fuels, mainly coal.

The variability of most renewable electricity technologies means that these technologies offer limited opportunities for replacing coal generation, and these opportunities vary across the United States. Given that the best wind and solar resources are located far from population centers, opportunities to use wind power and solar power are likely dependent on development of transmission lines. Without major breakthroughs in technology efficiency or energy storage, growth of power generation from all renewable sources[59] is expected to be limited to no more than 16% of all generation by 2035.[60]

---

[55] "New coal-, coal refuse-, oil- and petroleum coke-fired boilers and IGCC [Integrated Gasification Combined Cycle] units should also be able to meet this standard by employing carbon capture and storage (CCS) technology." Ibid.

[56] Diane Kearney, *Coal Projections from the Annual Energy Outlook 2011 Early Release*, EIA, April 6, 2011, http://www.stb.dot.gov/stb/docs/RETAC/2011/April/EIA%20annual%20energy%20outlook%202011.pdf.

[57] Total levelized system cost. "Levelized cost is often cited as a convenient summary measure of the overall competitiveness of different generating technologies. It represents the per-kilowatthour cost (in real dollars) of building and operating a generating plant over an assumed financial life and duty cycle. Key inputs to calculating levelized costs include overnight capital costs, fuel costs, fixed and variable operations and maintenance (O&M) costs, financing costs, and an assumed utilization rate for each plant type. The importance of the factors varies among the technologies." See EIA, *Levelized Cost of New Generation Resources in the Annual Energy Outlook 2012*, Annual Energy Outlook 2012, July 12, 2012, http://www.eia.gov/forecasts/aeo/electricity_generation.cfm. (EIA NewGen)

[58] See CRS Report R42532, *Carbon Capture and Sequestration (CCS): A Primer*, by Peter Folger, and CRS Report R42496, *Carbon Capture and Sequestration: Research, Development, and Demonstration at the U.S. Department of Energy*, by Peter Folger.

[59] Includes conventional hydroelectric, geothermal, wood, wood waste, all municipal waste, landfill gas, other biomass, solar, and wind power.

[60] EIA, *Annual Energy Outlook 2012, Table A9*, DOE/EIA-0383(2012), May 2012, http://www.eia.gov/forecasts/aeo/ (continued...)

Since load must be balanced on a continuous basis, units whose output can be varied to follow demand (dispatchable technologies) generally have more value to a system than less flexible units (non-dispatchable technologies) or those whose operation is tied to the availability of an intermittent resource.[61]

Development of joint renewable electricity and natural gas-fired power projects as "hybrid generation" is a possibility in certain regions. This could make variable renewable resources like wind and solar power more dispatchable, and would likely be dependent on state renewable portfolio standard[62] requirements.

Nuclear power may be able to take over more of the base load generation role from coal, but the path forward for new nuclear generating units in such a role is uncertain. According to the Nuclear Energy Institute, as of 2012, five new nuclear power units are under construction, and maintaining nuclear energy's current share of 20% of total generation would require "building one reactor every year starting in 2016, or 20 to 35 new units by 2035."[63] The cost of building new nuclear power units may present a barrier to this goal,[64] as might concerns raised by the nuclear incident at Fukushima.

Given current technological options for power generation in the United States, the long-term expectation of lower natural gas prices in combination with existing and proposed EPA rules suggests that the option of coal use as a fuel for electric power generation faces an uncertain future.

Industrial Coal Consumption[65]

In 2011 approximately 72 Million Short Tons (MST) of coal were used by the U.S industrial sector, which represented roughly 7% of total coal consumption in the country.[66] The industrial sector's share of U.S. coal demand has declined substantially since 1949, when the industrial sector accounted for nearly 45% of annual coal consumption (see **Figure 7**).[67] The decline of the industrial sector's share of coal consumption is generally a result of three developments since 1949: (1) reduced coal use for coke production and steel making, (2) reduced coal use for industrial process applications, and (3) increased coal use for electric power production, which now represents more than 90% of U.S. coal demand.[68]

(...continued)

pdf/0383%282012%29.pdf.EIA, Table A9, Electric Generating Capacity. Annual Energy Outlook.

[61] EIA NewGen.

[62] Renewable Portfolio Standards are state programs requiring a percent of energy sales or installed capacity to come from renewable resources.

[63] Nuclear Energy Institute, *New Nuclear Energy Facilities*, 2012, http://www.nei.org/keyissues/newnuclearplants/.

[64] The two units being built by the Southern Company at the Vogtle station in Georgia were originally estimated to cost $14 billion. See Ray Henry, "Building Costs Rise at U.S. Nuclear Sites," *SouthCoast Today.com*, July 15, 2012, http://www.southcoasttoday.com/apps/pbcs.dll/article?AID=/20120715/NEWS/207150345/-1/NEWSMAP.

[65] This section was written by Phillip Brown.

[66] Energy Information Administration, *Annual Energy Review*, September 27, 2012.

[67] Ibid.

[68] Ibid.

Figure 7. Industrial Sector Share of Total U.S. Coal Consumption

(1949-2011)

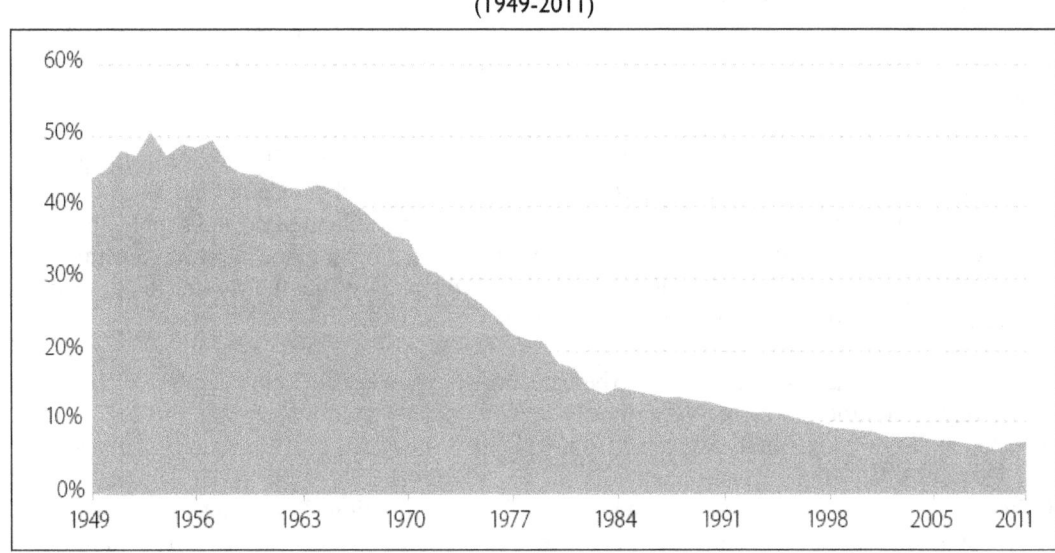

Source: Energy Information Administration.

Coal has a number of industrial uses including steel making, cement production, combined heat and power (CHP), and process heating. Additionally, coal is used by a number of industrial operations including alumina refineries, chemical producers, pharmaceutical companies, and paper manufacturers.[69] According to the World Coal Institute, there are thousands of products that include coal or coal by-products: soap, aspirins, solvents, dyes, plastics, fibers, and specialty products (e.g., activated carbon, carbon fiber, and silicon metal).[70]

The Energy Information Administration classifies industrial sector coal consumption into three separate categories: (1) Coke Plants, (2) Combined Heat and Power (CHP), and (3) Non-CHP. **Table 1** indicates the amount of coal consumed in 2011 for each of these categories.

Table 1. 2011 Industrial Sector Coal Consumption

(Million Short Tons)

Coke Plants	CHP	Non-CHP	Total
21.4	24.7	25.6	71.7

Source: Energy Information Administration.

Issues for Congress

Given the expected increase in natural gas supplies and expected shift in electricity production from coal to natural gas as a primary fuel, the issue of what level of coal generation should be preserved for power generation in the United States may be a question Congress will be faced

---

[69] "The Coal Resource: A Comprehensive Overview of Coal," World Coal Institute, 2009.

[70] Ibid.

with. While U.S. coal currently used for power generation may have options for markets overseas, the domestic future of coal power generation is clouded by both market prices and environmental regulations. The United States is believed to have a bountiful supply of coal. But how much or whether the future of coal will include power generation may be a question of national energy policy.

Congress may be also confronted with the question of whether an "all-of-the above" strategy for energy is viable. The electric utility industry values diversity in fuel choice options since reliance on one fuel or technology can leave electricity producers vulnerable to price and supply volatility. However, an "inverse relationship" is developing for coal vs. natural gas as a power generation choice based on market economics alone, and policies which allow one fuel source to dominate may come to the detriment of the other.

Congress may also face a decision as to whether coal-fired power generation can be a viable technology option for the future. New advanced NGCC plants entering the market in 2017 are expected to cost approximately $63 per MWh, while new advanced coal plants are estimated to cost $111 per MWh, and advanced coal with CCS is estimated at $139 per MWh. The cost difference between these technologies, and proposed GHG emissions limits, raise concerns over coal's future as a viable fuel choice for power generation. It is not clear (at this point in time) whether research and development will result in new technologies able to reduce the differential.

Author Contact Information

Richard J. Campbell
Specialist in Energy Policy
rcampbell@crs.loc.gov, 7-7905

Phillip Brown
Specialist in Energy Policy
pbrown@crs.loc.gov, 7-7386

Peter Folger
Specialist in Energy and Natural Resources Policy
pfolger@crs.loc.gov, 7-1517

Acknowledgments

James Uzel and Amber Wilhelm contributed to the graphics of this report